For Wilka
and for Kim

Thanks
[signature]

OUT OF NOWHERE,
THE BODY'S SHAPE

OUT OF NOWHERE, THE BODY'S SHAPE

by

Beth Simon

Pecan Grove Press
1 Camino Santa Maria
San Antonio, Texas 78228-8608

Cover Photograph by Carl Yoder

Cover design by J. A. Hilderbran

ISBN: 1-877603-32-5

Pecan Grove Press

Box AL
1 Camino Santa Maria
San Antonio, Texas
78228-8608

(201) 436-3441

for Tom

Acknowledgements

These pieces, some in other forms, have appeared or are forthcoming in the following publications:

Alaska Quarterly Review, Southern Indiana Review, American Literary Review, Arc, Calliope, Iowa Woman, The MacGuffin, Midwest Quarterly, Nimrod, Northeast, the Seattle Review, TriQuarterly, Wisconsin Academy Review, Wisconsin Review

I've never longed for the annulments of heaven

—Lynn Emanuel

Contents

Out Of Nowhere, The Body's Shape

How It Looks From Above

Balance, Then Rise

OUT OF NOWHERE,
THE BODY'S SHAPE

the huge uncertain water of the heart
—Erin Moure

SAP

Good Friday, the conifers are rich in resin. A cemetery
near my house hides a stand of long leaf and spruce.

Sly young girl, I sneak out of school, go home, make myself clean.
I snag the old man's raveling cardigan off the back porch nail,
wrap up like a secret, slip out through the root cellar, past
the crabapple full of jays who suddenly cry my name.
I enter the rutty field.

Lost in frost fog, head down in the dead brass of bean waste,
dry wind fills my ear. My feet grow hard, cold, heavy, they rise,
fall, rise on chalky furrows, jabbed by sharp-edged clods thrown
in fall plowing.

Then, I'm across. Up the small hill. Emerge, sweaty,
blown, at the graves. At the tree, the white pine, its harp
body dark in thin sunlight. The lower branches swirl
ground and I crawl under.

Dim. A held breath of cinnamon and rot. Around the base, a rug
of needles, and I lie down, face to them, smooth as rice. I roll
to one side and clutch a low bough. I trace its silk across
my cheek. Again, then again, then curl to the trunk.

An amber rill has flowed down a crevice, congealed on a knee
above the base. I press against the bark like earth like mint.
Lick. Nothing. I jab with my tongue, scrape pitch onto my teeth,
and it is cold. Rank.

Such bitterness. My jaws snap shut. My mouth swells with saliva.
I cannot swallow, cannot spit. Vegetal, animate, a raw unfamiliar
life. Here is the heart.

MILWAUKEE ROAD

Right on schedule, July pulls into the Des Moines depot
long enough to load me on the Milwaukee Road and lunge
out, a lung-clogged *chug chug chug,* that is my parents
getting rid of me, rid of me, shot north, a bullet

defining the horizon of my age. At nine, ten, eleven,
I am a premonition with eight hours to live out
everything between my dreams, Afghani escape,
Arctic survival, the arrival of love, until somewhere

in a new quadrant on the bare map of my mind, we pick up
velocity like a hitchhiker whose swift intention
I know as well as my face. Take me, I breathe, clinching him
hard, hurtling into the slick of possibility.

I LEARN TO FLY

Once the trains are shaky enough even my parents admit
I might not be there to collect at the other end,
they produce the miracle, airfare and wherewithal,
for the skycap, for the attendant nearest the door,
book bag, sci fi, command to Look out.

I don't need them anymore. We're skating down
the runway, first kid in the mob, hiccough
off the tarmac, nose the sky for scent,
once around the airport block, and my heart,
dumb bird insisting north in high wind.

BRINJAL

Sixth grade spring, front row, first desk on the right,
the exchange girl gave us such joy. We discovered
velar clicks, green and sharp in her throat, the breathy
blossoms of her English. Her firm instruction
not to say 'Rhodesia.'

She taught what love was possible in hard Iowa
heat, ninety on April tenth. We wore high-top sneaks,
torn t-shirt, boxers, she, thongs, sarong, bandeau,
until, breathless on the edge of new land, we learned to simply
embrace water. Oh love, even now, let me dive deep,

grab hold and swing round and round the cool blue curve
of your back like the playground pole, wire myself in the fine mesh
of your hair, and pulse florescent beneath your surface, silky
as sea grass, outspread as kelp, mineral rich, nearly ripe,
safe from the alkaline air of division, of past tense.

GRAVITY

On a ledge above Bryce Canyon
I watched my sister in morning sunlight
unfold herself, butterfly back muscle,
arms, a luxurious harp over her head,
clear-tendoned as Gabriel,
she called me up, north,
toward jet tails twenty thousand feet high.
Then she simply stepped off
on nothing, back lit,
shimmering, a moth pinned
in almost perfect stillness
outside time, sculling on filament and dust
and separating clouds like thin silk
and on the next beat, as if the single thought
that held her had turned to something new,
she arced and fell one moment from heaven.

So. What am I to do?
Skulk across mid-air playing Joshua?
Rewind the sun? Make all time stop?
How even mark this trajectory?
A hundred foot of rope draws nothing
but a rising cloud of dust
and loose shale rattling the cliff-face.

GROUND LEVEL

That summer, mist rolled down the valley
like bad luck, a snake shedding skin
after skin in its single thought
of the entrance to the cemetery road.
All night parked cars rocked and moaned
around the Stone Angel. That was sex.

I am not eleven anymore, and have entered my own
August under thunderheads like glaciers.
Back then, I did what I wanted and yet
they all forgave me as they would
an annoying child. I forgive nothing.
That summer, people said I lost

my sister, as if I didn't know exactly
where she was. Any day I want, I go to her
grave, bring her out, saying, "You stand on this
boneshell surface nothing like earth. You trace
these dates while light fades." And then we imagine
the dark, later, the underhand arc of the thin-armed

boy, his eagle tattoo, salt taste,
how he breathes into my blouse until
my fingers slip his leather belt. Tonight,
his stolen car, The Angel. She loves to feel
herself returning to this place. She mouths at him
I love you when we turn away.

OUT OF NOWHERE, THE BODY'S SHAPE

When I was ten, eleven, twelve, I lived
Saturdays in a movie theatre balcony
front row, sneakers kicking like a clock
over the coke-slimed floor, swallowing

chocolate covered anything, chin on guard-
rail, aimed at The Sundance Kid, Super Fly,
The Graduate, hours spent kissing upper lip
against lower, learning the few forms of love.

Those filtered afternoons swam straight to the dark's
first beach, mineral sand warming the body's shape
out of nowhere into sea plums, astonished
that the boy touched me there, inside. The revelation

that he was a simple series of waves—
blood high seeking, eyes unveiling
new moon whites, dolphin heart skimming
the surface of the water turned silver

under the rock cliff, salt voice begging Mary,
Mother, how modeling the way his lips
made safe harbor, the fluting of his hands
like doves died down taking refuge in the night.

CARIBE

I was sixteen, warming up in the evening waters of Tobago,
Trinidad, the whole reef of Caribbean possibilities
luminous behind the junk Dodge & the wire thin boy kissing
sudden lips at me from the dark of the unrolled car window, such
small wet sounds to pull me to his stolen back seat like a magnet.

This was swimming, the green ribbons of clothing, the sweet bad
liquor he licked from my thighs. His hands, quick sand fish,
dissolved into moon white shallows that made our own shoreline, and
just up the beach, the resort, the buffet, the big band beating
big metal drums, soprano sax, marimba, rhumba dance floor
brilliant orange, green, red, everybody on fire.

HISTORY

Grandmother, at ninety-three, still
head wrapped in braids, pours out
tea from the pot I've known all my life.

She says, "Listen.
My father was a big
big man. You worked for him,
he paid you cash.
He didn't ask twice.
Parts of this town
shook under his boot.

(She makes her fist a kettle,
thumb, a spout, smacks the wood table
hard)
Those days
He went in the pony cart
with me. Out in the back lanes
Hunks drank
Czechs baked
Jews banked,
And everyone knew who he was.
He could have owned anyone.
A natural Democrat.

Now listen. There's more.
Father came home
from The War
with no leg,
but what did he care?
He brought china from France.
Each night of my life, he drank
where you sit, sucked

tea through a sugar cube
gripped in his teeth,
the color of tea from this pot.
I see him just
as if he were here, my God,
big man. I've promised
to never forget,
each cup is his cup."

MY FATHER'S LOVE SONG

A night like my father's favorite
for the long drive to Sec Taylor Stadium,
for the Des Moines Demons, a farm club
we thought was Triple A.

Pushed back from supper, we slide off to let
him kiss Mom like a man who has time.
At the car, he pats for jawbreakers,
smokes, hunting while we kids ease the backseat door,
stealing in on our knees. He barks, "Watch it"
and guns his Buick out the driveway.

I love baseball.
Dark air, strange bodies,
squirrely on root beer and Hamms,
changeup wild in the dirt, arms
wheeling, Sal Bando rounding third
as if we're all heading for home.
My brother says he saw some great ones pitch
but I was safe in my land under the bleachers,
singing and digging in dead grass
and catnip and ripped plastic cups,
making tracks for China or Mars,
easy in the ocean of broad sweet curses and moaning
spread above me like the wash of new birth.

My father never forgot me.
In the seventh inning stretch
he'd park my brother at the cap stand, then inch
along the baseline crooning tenor through the slats,
"Sweetheart, oh baby come to poppa."

MY SISTER IN A SNAPSHOT

At the street end of the driveway, flash smile,
hard squint, front tooth gone, the perfect
relation of fat to muscle to skin,

backpack behind her leg like a child, these say
"Don't need." "Leave." "New life."
What can I know about her, my sister, mine,

who drowned the summer I first tasted beer?
Up from Shonkin, floods rush valleys like wind, blowing out
pole buildings, wheat, trailers anchored in steel,

what do I know of the water line, how they found her
wedged against rock, hair mixed with shale,
how she came there tracking lambs, took a wrong turn

west, stopped, in a deadend canyon, entranced
by the firestorm sky. I say she died for me, called
my name within the hollow of her hand, sent ahead one

half-live echo circumscribed by season, marked
in bone this one grey trace shading north, high
plains, that lost direction no map can lead me in.

BEFORE AND AFTER THE LAKE BIRDS

Before he forgot
everything, my father drove
a nail in the window-side wall of his one-room
supervised apartment and fixed
the first of one hundred
photographs of my mother.
He wanted to record what he meant
to be memory, to set
each feature on a still,
present surface.
He was that much in love.

This first photo shows her on
a grassy spit at night. The moon
must be full, over water, coppering
boulders several feet out from the shore of a lake.
On a day he knew me I asked
what lake.
He thought maybe Superior,
Michigan side. He didn't remember
exactly when he stood on this verge
or his age.
"Sweetheart," he laughed,
touched my face,
"I was young."

Beyond my mother in early shallows,
darker rock, mica-shot, breaks clearing water.
On the east boulder, a tern, or it could be
a gull, waits for its moment,
a slight shift of stars,
a freshening wind,
to raise wing as an echo.

HOW IT LOOKS FROM ABOVE

But there is a place that will not change,
a place that is rooted in dream
—Larry Levis

NOON IN THE NORTH

The bazaar smells of straw,
excrement, sweat.
Rickshawallahs cadging drinks
at the Parsee liquor stand
lean away from the seized up ice box
where I hide with shy women
and pray the sun down.
The ride to the river
is too far to ever begin.

I have learned to distinguish
clean dirt from filth,
rats with disease,
bad fruit, old tea,
the damp kiss of meat.
I accept the cracked
glass weeping sugar cane juice,
avoid the chipped edge,
savor the bite,
swallow and smile,
nursing my teeth.
Hand the glass back politely,
never touch flesh.

BY SUICIDE, BY DROWNING

You swore no ritual and now,
look. Laid out, glass-eyed.
What gall. I'd call it chutzpa
if I weren't too angry,
if your mother weren't here,
if I knew how you farrowed
each salt-sown kiss.

And what, after all the furor,
has changed? Every lopsided fight
after love ended naked,
back lit in the icebox light,
skin against bone, gripping the long
cold necks of brown beer bottles.
I turn, not much, and see

your curved back, rich worm of earth,
half bedded in sea routes dug
while I slept. You coward.
Let them bury you.
Grief is my pocket of dark air
in dark water without surface.

FROM A SMALL BOAT ON THE SURFACE OF SUPERIOR

The arms, telegrams held for delivery,
attach under the uniform to the still muscled torso in strips.
Without boredom, agape in his motion of God, face up,
if the lake surface is up, legs easing softly apart, together,
year after year, the freighter mid-shipman recalls
a crucifix, 60 feet down, 30 miles off Thunder Bay,

gentle after his long swim to bless the silver flickers
at his wayside. The hatches stay open, dinner abundant,
a watery portion that never goes bad. It is far too romantic
to give a dead man wife or dreams. No species of interior bone
and flesh releases itself back to daylight form.

HOW IT LOOKS FROM ABOVE

I know the drunk in the doorway
like a sack of potatoes half out
of the rain, humped under the bulb
in spilled coffee, wet scrap.
I see urine. Not his face.

Here is night, the father camouflaged,
muttering grievance, slouch hat covering
visible shake, the years since he yelled
"Get," and didn't mean it when I slammed the door
against him, never knew if he'd unfisted,
bowed his neck, called me home.

This one has easier needs.
He has jammed himself into a place
no wider than the weathered bricks,
moaning loud enough.
He has matched his cough to my stride,
and though I keep my hands in my pockets,
he can tell which holds keys,
and which change.

If I could rise above this city,
containing within my sight the shape of black
roots and refuse and drunks, all questions resolve.
This is a sack
in the doorway of my Goodwill,
gradually soaking up rain.
Simple and still, like sloughed skin,
water refracts street light.
Stars dissolve into stone.

BRIAN EXPLAINS HIS ANNIVERSARY

I admit that certain sounds do it.
War shows. Noise in the kitchen.
Pan banging. Door slam that starts
that beagle. No, but this time, I'm sure.
Trouble. Someone pushed too far, door ajar,
dog escaped, something, because how else account
for the big Buick, just new enough,
parked one house down, right fender's wrong blue,
and bent. That high rev idle?
You've got to ask,
I mean, this late in the day,
no one's warming up, though,
you're right, it's early, no one's late.
But still. Next door shades move
when they think I'm not looking.

Well, I am looking. I see things
move, beg, waver on the perimeter, and
flight even yet isn't out of the question,
because there's that noise,
copter blades, *Whomp, Whomp,*
I mean, just tell me.
What do you think more likely:

The whirlwind, from nowhere, Whack!
and overhead, the whole air,
world turned green, and praying, Please,
God, Rescue, and maybe the chopper
does clear the hill, a rope, line,
but they're full up, or they want
your buddy, or even if you catch,
your best hope is being yanked skyward
like scrap. That's one possibility.
But what if there is no Just God,
only me? And I'm alone in the house
with the windows open
next to the one with the TV on,
where the neighbors are gone,
and the young pup begging to go out.
Wouldn't that prove something,
bring the wind down to a breeze,
a simple quiet where a man can think
it out? Wouldn't it make you weep?

DRIVE

Heat-stoked will like wire,
I burn I-90–
two hundred miles to blow

you out of me. Miss
the exit, right front tire
kicks rock, car slews

jumps the ditch, half rolls,
hatch pops, I fall
unborn out through steel,

oiled smoke, on asphalt,
blood or glass & broken
mirrors & weeds on fire.

A GLOW ON

Even in the least moonlight you can see
the trace of what went before,
"And we have lift-off," she says,
her arm raising
the empty bottle from the pilfered rug
beside the sofa where she lies
out. That bottle next
to the second one. She always
has a beer in the evening.
A shot between beers.
A couple of beers.

First, she calls.
"Come over. I've made
spicy things. Saffron rice, spanish sauce,
lime. You bring the wine." You manage
small sour sips like child kisses.

And on the sofa
pearls. Her breastbone a ship
under the skin, an inland sea,
her skin a map.

Oh, she calls,
"Come back. I'm stirring
curry, chilies. I want to see you
in dusk. Wear nothing
under your coat I can't remove by wishing."

You finish the wine,
body like almonds, dark blood,
translucent and beating
red ruby red.

Those nights open
"To love," you say. You're at the point
to toast anything. Pour one glass
and she pours one, drinks half, fills half, drinks half.
She can drink a bottle,

Everyone has a spot
after dinner, throwing hard
cover books, salad, oil like a net
of stars in your hair, like starfish
on the wall, glass shatters glass,
the air, full of dirty words.
Liar. Thief. Eviction. Fix it.

Breath or silence or her
loose turning, as if breast,
arm, disparate. She tries
to rise, says Get out of my way, and the lights
are on. Left is counting.

LIKE PENUMBRA

You want
words like penumbra, hibiscus,
blue in the awkward position,
or sweat or gold, like sex,
which you don't mention, or diamond-
hard earth or mica, yellow on
silver like urine.

Denial, a dark
aroma adorns you, vicarious, mossy
as conifer grown deep at ground root,
bedded in green needle musk, circling
branches, caressing wild shoots,
silk knotted like vine.

Surrounded by stripling
white birch, now vicious, like axe or tomorrow,
clean slicing, translucent, denying the root.
You adore it.

BREAKING GROUND

Even as I wake, pack, burn
toast, you rule out back,
Northrup King of fecund weather,
drilling the garden for gold strike
melon, high stake pole beans,
parades of feathered fennel waving above
cayenne like ostrich earrings.
You root in scab wood, blocked light,
dry afternoon.

Last night I left all catalogs
on your desk, all seed promises
of sweet ground water.
Last year we had drought.
Salt dirt, bitter as gourd,
squash swollen with white worms,
the eroding rotatation of weed,
hoe, plead. You have a need to fruit
I haven't got anymore.

EARTH AND HEAVEN

The air is ripe, plum wine, and I am worse
than drunk on it, eyes open, in the passing lane,
shooting past the child-clogged Nova,
the LAF Colt with the bashed in hatch
jigging like a snare drum in a jazz club,
all of it, blur, but awake now,
terrified and scrupulous enough
to manage to pull it all back into line.

I know everything. The highway patrol
camouflaged east of Valparaiso is nothing new to me.
Three times a year all my life,
I drive this line home. Today, call it
"Indiana Toll Road," breasting Gary
through smoke, dreamy pale rise
into smoke like bedsprings of snow
from what seems archaic marsh or swamp,
the final edges of Michigan, shooting
the skyway bridge into blue sky way,
all this by legal decree that my mother
and I take two sides to everything.

What else am I waiting for but You
to part the fat cumulus clouds like smart sheep,
or drapes, smoke, bottomless as revelation.

Oh, let there be The Judicial Choir,
robed and omniscient, Valkyrian bellowing:
"Your mother is wrong. You have suffered, endured.
You are better than a saint and rightfully aggrieved.
At the threshold of your mother's house you shall
be justified, and in the final end,
you shall be redeemed," hysterical
and flawlessly tuned to some ageless
Brahmanical cantata that if I know anything at all,
I know this is mine, just this once.

BALANCE, THEN RISE

Can this still turn out to be a love poem?
Can I still pull you from the wreckage
—Stephen Dunn

BANARAS WITHOUT RAIN

Coming down from green mountains in August
after monsoon, I heard that all rains
had failed for the third year in a row. Yet, here,

in Banaras, we wake to high thunder,
lime fragrance above dust,
milled wheat blowing from Shri Dev's shed,
the Patak brothers shyly searching
each other for nits on string cots
near the road;

this long even light of six, seven o'clock,
the hard red surface of Durga Ji Temple
gleaming, nine hundred years old,
the blind Muslim seer, knotted in rags
blessing kohl-eyed girl-brides who offer him milk
sweets, barefoot in silence amid crows.

Here is beauty,
or chance, new, drawn from the night
out of patterns to red jeweled paths,
rich leaves, porphyry massing underfoot.
Yellow moon just rising.
Grasp of jasmine.
Blue mist.
Heat.

FEEDING BIRDS

Last night was your last free of tubes. You had been, as they say,
holding your own in a bed that still meant what it means for the living,

more than sleep, the embrace making easy the equinox passage,
that minute shift of season, letting even smooth muscles relax,

and you, reborn, riding the air, ski sideways then drop spine-first,
onto the simple sure surface of concrete.

You bled freely. Last night, three pints. Fifteen minutes left you
five minutes from death. No, what is dying? Yours, now, the window

the feeder, the aspen and you, awakened to newborn command take
up orioles,
jays, like fourteen-year-olds bolting third hour, hop, glint, and pump.

Reach out your left through bars that make a box of your bed. Here.
Never the miser, you scatter black seeds of the sunflower for all

that's left to endure. Psyllium for this hard November.
Thistle to bridle the wind.

KUGEL

When I up and left my husband, we were all shocked.
Years later, he told me that when we first met,
he knew. He said to himself, Midwest. Native.
He said, Fat and Sassy, here I come. He said,
I came out to Iowa to find you.

What could he have been thinking, my now exhusband,
either of us, to ignore the prominence of my cheekbones,
the simple definition of my ribs? The way my stem
quickens from my pelvis like a brief spice plant
in a thin clay pot.

We bought a house with ghost-riding attic,
incontinent water heater, a furnace we learned
to hate. A circle of ash trees like signal torches.
Each fall, our yard bloomed under migration while we,
sweet midnight, sought geese as they swam down
the north sky to cut corn.

Kugel in the city was my mother-in-law's dish,
delicious and rich, cinnamon, butter, baked until
the noodles on the pyrex edge crisped. This was the warmth
and the way of her welcome, a Sabbath thanksgiving
for an only son's return to New York.

But I, uneasy in midtown Manhattan, and strange to
new seasons, snaked myself back across alkaline plowfields.
I tracked by the stubble of granite pocked straw land,
glowing like moon light, pressed hard to the lee side
of silence and wind.

GIVING OVER

On the quietest day in June,
when the Ocooch hills
stand a little farther back,
the rosehips widen slowly,
and the wind
like a small fish
runs silver in aspen, a day
when air is nothing
but light
and the broadwing hawk
tips the flash of an oriole

I rise west above the blue ridge
Kickapoo flood plain,
I want to taste the root of
this very hill,
trace the creek's ice heart
to its source,
lie down in my bones
like white birch,
leave only clean limbs to the sky.

NORTH

Like trout, above Ottawa, in August, we are
bones, rising to surface too late for summer
in this hard river. Silver birch, old aspen
are shadows and we, too large and far whiter,
chart water over rock.

Spruce bears itself out of granite until, overhead,
geese rise south. Mosses burrow, entering chirt, leaving
ice, a membrane over earth, and the last, above us,
is dark, covering absence, reflecting earth, ice, stars.

DRY

Marko lies on the backseat talking
trash, smoking dope, all my cigarettes,
right arm bent to make
the ridge from his nipple
to hip a cold white creek. He lights the next
from the one he's got going
the way he's got me going.

I'm kissing him wherever.
Syl thinks Marko's no good but
that's so easy when she doesn't have to be
anything but the forecast of flash
flood on a dry hill. He lays out
what he wants, line by line. Marko is very articulate.

When he opens his eyes, hoping the dark
cloud spread over is the clean
rain of Syl and he the sown field of rye,
I work on him and dream
I'm Syl, cloud, the full weather of my breasts
hillsides of green, and when he opens
his eyes, I'm rain.

SPIRITS

She'd like to kiss it all
off tonight, the warm beer
she's only holding,
the nearly same band
with the same fuzz
pedal, reverb, offhand snare,
bass guitar player's pointed snake
boots aimed into her breastbone,
the way he's shaved bald
up both sides, shaking
his black-dyed pony tail,
how his silver ear clips
reflected in the moon
of tiny mirrors turning on the ceiling
sink their lure
taut under her belly skin
down about a quarter inch,
even as up top,
outspread against the white vault
the bat wakes
in the cool grey
silk sheathed
cells of the cerebrum,
thinking, Old Hat.
Last Call.
It's Been Done.

THE FORMATION OF BONES

The bad, bad brother of girls
I was afraid of in high school,
you smoked in dark bathrooms, kicked wide
the lunchroom doors, and when you popped a match,
became the cloud of sulphur.
I fell in love. You said
I thought I was too good,
your tongue a Goth against my teeth,
opening by signs everything
I dreamed you knew.

I wanted out of my skin.
From World History, I watched your gang
circle the parking lot. When you,
nose to palm, inhaled then dropped
dead weight, the rest, the grackles, sought
the county line, Coast Guard, jail,
well before the ambulance arrived.

I cut math for the River,
lay in long grass listening to sexless
women in men's shirts, bamboo poles
laced with trout flies, recite the liturgy
of surgeries and California
grandchildren. One lied
about the size of fish
that were never anywhere near Iowa.

I refuse to call anything
After. I deny you
parole, discharge, on-the-job training.

I will not allow you to kneel in your driveway,
Marlboro between your lips,
while you mend your daughter's doll.
These days even the tattooed manage
to own a house, yard littered with bikes.
I do not give them to you. No, now
I say your hands are sticks.

SMOKE AND ASH

My hairdresser smokes.
They all do, waiting.
Every six weeks, thick snakes
through my scalp, mango shampoo
in the wet smell of filters.
He gives fast moving haircuts,
half inch and a blow dry,
says, "My boy's dyslexic."
"I've been straight for three weeks."
Weekends don't count.

Govind cuts hair in Bombay.
Chain smoking, chanting,
"Beauty, be Hindu,"
he taps me with almond oil,
white fire curls from his thumbs,
dead vine, burnt wood, rock.
His breath is mimosa,
his lips are shy insects saying
Marry me marry me take me to Tucson.

His room is clogged, fat gods in dirt.
My hairdresser's waist is acrid and sweet
a lotus in my hands, and I long to try it
but I hate heat, mirage, desert sand.

PREPARING THE ICE AGE

Stroke waits on my future like my own private bridegroom.
You want Truth? Skin withers, also secret
desires to smoke, get blotto, pick a bleak fight,
receive absolution, those cheap stolen goods
to unplug the phone, then flick off the switch.
I avoid difficult meetings, dinner with friends,
fear dust, even in such a wide and pliant room.

Unsettled valleys blow with husks and cold notes.
Night air booms, swallowing dry sand frogs and wind,
and these are but metaphors for transmutation, or winter,
yet I can't hear above the hustle of leaves, of bone
oxidizing, by laws of genetics into calcium spars,
syncopated heartbeat, residual scar tissue,
embittered tendons, cortical crashes, the splintering
fanfare as glaciers break loose to retake the south lands.

BALANCE, THEN RISE

Rowan Creek Sanctuary
abuts a shooting range.
Easter Sunday, air is
gun-filled. PingPingPing BOOM.
Grackles pump skyward, blossom,
drawing blue bitter smoke.

We fan out beneath endless clouds,
edging Marle pond
between red oak and fir.
Down to earth, skunk cabbage,
blood secret, opening one veined cup
crowns moss.

Above us like glaciers,
thunderheads shift west
and in that breath
between pistol and pistol
light refracted, a prism,
two phoebes brush cat tails,
a heron lifts,
a sharp-tailed sparrow.
Then rifles beat the rush of wings
back into dark pine.

Yet even as the silk grey
feather drifts to dust the pool,
the newborn darner taps froth.
Mica-shot feldspar breaks clearing water.
The marsh hen flies.

TOM IN THE TREES

Fifty straight feet up the ice-slick walnut,
wind chill twenty degrees below ten degrees below zero,
half sole of one shoe in contact with the last feeble piton,

silver bullet or button that wouldn't support an orphan
and that you only drilled to appease me, you are reborn,
with three right fingers barely pressing the bark to anchor

my flying dream. The rest of you unfolds toward branches even higher.
From the kitchen, I see you tremble in time with the hardrock
grunge band in your headset, and your left hand, a light-limbed

angel, raises the chain-drive dead weight. Saws blue air. You are
disarming the forest for my twenty dollar order from Johnny's Seed.
My little questions, Do we get enough sun for tomatoes?

Lettuce? urged this terrific activity, where every rear
window that gives onto our back yard reveals a wild-eyed woman
with a phone, on guard, ready to do, what, call 911?

The next door neighbor, who raised twin bikers complete with
Harleys, Dobermans, knife fights, cops, stands exactly beneath,
would she break your fall? screams "Get down from there. You are giving me

a heart attack," but you have entered the snowline heaven of height,
a Himalayan paradise above redemption, your reaching heart, against
all odds, braced to vault the arced frozen mountain of your ribs.

ABOVE OCOOCH MOUNTAIN

You lay dry oak in the fire ring,
I, the buffalo blanket on grass.
We lie down, earth nervouse, and wait,
jumpy at bats, high jets,
the howling of farm collies craving sheep.
We will the sky to darken
the night to rise,
sharpen, bloom.
Light lasts in July.

Above us things move.
Orion rolls deep
into laced aspen arms,
Arcturus dies hot on the moon's
arcing breast, and yet,
how shall we, without need,
convection, some law
like raw heat
draw to heart, bone, breath,
quicken blue flame up
from low glowing ash,
break like a comet
one last skin of stars?

SEMPER ETC.

Each morning at six I watch the man come by
with the dog. He reconnoiters my side
of the block and releases what is called
a chocolate lab, who noses into my neighbor's
bridal veil and unloads. I suspect them
of waiting for light when such acts make
relief for us all.

The man keeps the dog on a short lead
and the dog is as parallel to the ground,
his legs as strident and clear-tendoned as the man
is upright. They have patrol posture.
I think they met in the Marines, pumped iron,
marked enemies, crawled, muzzle by cheek in mud
on maneuvers. Bound now, traveling by blood,

they walk their round as I guard my window,
frame to frame, the three of us on a duty we shoulder
at reveille. The sun is up. The dog's bristle coat
glints a shade or two darker than strict regulation
yet well within range of the man's gold parade stripes,
flat-top, waterproof, lace-up boots, the orders by which
our path is drilled through the uniform to the vows.

Books in the Pecan Grove Press *Signature Series:*

Somewhere Between Hither and Yearn
by Ritalinda D'Andrea

A Time to Be Born
by Olga Samples Davis

The Skies Here
by Rachel Barenblat

Dragonfly
by Vince Gotera

A Measured Response
edited by H. Palmer Hall

Snow in South Texas
by Cynthia Harper

A Certain Attitude
edited by Laura Kennelly

Choc'lit Milk and Whiskey Voices
by Jo LeCoeur

Pointed Home: A Cross-Country Essay
by W. Scott Olsen

At the Border: Winter Lights
by Carol Coffee Reposa

Out of Nowhere, the Body's Shape
by Beth Simon

Stone Garden
by Barbara Stanush

And forthcoming in spring 1996:
Counting Out the Millenia by John Oughton